THE EQUATION DILEMMA

Terrence Howard's Critique of Modern Mathematics

Oteren.fredrick

DISCLAIMER

The viewpoints and sentiments conveyed in this book are those of the essayist and Terrence Howard and don't be ensured to reflect the power methodology or position of any educational establishment, affiliation, or others referred to. While each work has been made to ensure the accuracy of the information presented, the maker and distributer make no depictions or certifications as for the satisfaction, precision, or suitability of this material for a particular explanation.

COPYRIGHT

© 2024 [Oteren.Fredrick]. Protected by copyright law.

No piece of this distribution might be replicated, appropriated, or sent in any structure or using any and all means, including copying, recording, or other electronic or mechanical strategies, without the earlier composed consent of the distributer, with the exception of brief citations typified in basic surveys and certain other noncommercial purposes allowed by intellectual property regulation. For authorization demands, keep in touch with the distributer, tended to "Consideration: Consents Organizer," at the location beneath.

TABLE OF CONTENT

INTRODUCTION
CHAPTER 1:
- The Foundation of Mathematics
- Howard's Underlying Experience with Arithmetic

CHAPTER 2:
- Identifying the Flaws
- Explicit Regions Howard Views as Hazardous
- The More extensive Ramifications of Howard's Evaluate

CHAPTER 3:
- The Issue with Numbers
- Explore of the Continuous Numeric Depictions and Assessments
- Elective Viewpoints and Approaches
- Genuine Repercussions

CHAPTER 4:
- Geometry and Spatial Understanding
- Howard's Reactions of Customary Math
- Elective Ways to deal with Math

CHAPTER 5:
- Algebra and Equations
- Howard's Reactions of Customary Polynomial math
- Elective Ways to deal with Polynomial math
- The More extensive Ramifications of Howard's Scrutinize

CHAPTER 6:
- The Role of Mathematical Education
- The More extensive Ramifications of Howard's Investigate

CHAPTER 7:
- Mathematical Philosophy
- Howard's Reactions of Customary Numerical Way of thinking

CHAPTER 8:
- Real-World Implications
- Howard's Reactions of This present reality Use of Arithmetic
- The More extensive Ramifications of Howard's Points of view

CHAPTER 9:

- Responses from the Mathematical Community
- The Eventual fate of Arithmetic Considering Howard's Studies

CONCLUSION
ACKNOWLEDGEMENTS

INTRODUCTION

Blueprint of Terrence Howard's Insight
Terrence Howard is a well known performer, singer, and producer, generally famous for his parts in broadly lauded films, for instance, "Crash," "Hustle and Stream," and the television series "Domain." Past his advancement in news sources, Howard has a huge interest in science and math, which has much of the time been obscured by his Hollywood persona. His advantage with numbers and conditions has driven him to dive significantly into the underpinnings of mathematical speculations, beginning both discussion and interest inside educational circles.

Justification behind the Book
"The Condition Trouble: Terrence Howard's Examine of Current Mathematics" plans to uncover knowledge into Howard's uncommon perspective on the field of math. Unlike ordinary mathematicians, Howard pushes toward the subject with a new and regularly offbeat

viewpoint. This book will research his investigates, examining the authenticity and repercussions of his conflicts. Hence, it attempts to open a talk about the significant guidelines that help present day math and the potential for reexamining these thoughts in creative ways.

Why Howard's Perspective is Intriguing and Critical

Terrence Howard's pieces of information into math are basic in light of numerous variables. Without skipping a beat, as an untouchable to the educational field, he conveys an other point of convergence to the discipline, unlimited by spread out principles and tendencies. His technique is laid out in a genuine interest and a yearning to understand the universe at a focal level, which resonates with various who feel alienated by the multifaceted design and impression of standard math.

Moreover, Howard's examine isn't just theoretical yet also significantly private. His encounters with mathematical thoughts have shaped his viewpoint, and his excitement for the subject is clear in his tenacious mission for clarity and truth. By examining math through Howard's eyes, we gain another perspective that moves us to examine our doubts and research extra open doors.

Finally, Howard's effect connects past academic world into standard society. His indisputable quality as a notable individual gives an exceptional stage to zero in on mathematical issues that could some way or another stay bound to savvy conversation. This book intends to utilize his celebrity status to interface with a greater group in huge discussions about the possible destiny of science.

In the segments that follow, we will jump into the specifics of Howard's assess, exploring the locales he acknowledges are deficient and his thoughts for improvement. By understanding his perspective, we can get critical pieces of information that may ultimately add to the advancement of mathematical thought.

CHAPTER 1:

The Foundation of Mathematics

Verifiable Setting

Science is in many cases hailed as the most flawless type of human rationale, a discipline that rises above societies, dialects, and periods. Its establishments were laid in antiquated human advancements, with early records of numerical idea tracked down in the vestiges of Mesopotamia, Egypt, and the Indus Valley. The Babylonians are credited with creating quite possibly the earliest mathematical framework around 3000 BCE, utilizing a base-60 (sexagesimal) framework that impacts our time-keeping strategies today. At the same time, the old Egyptians utilized science for functional purposes, for example, building pyramids and overseeing horticultural cycles.

The Greeks later reformed arithmetic by presenting thorough insightful thinking and conceptual reasoning. Figures like Pythagoras, Euclid, and Archimedes made critical commitments, formalizing calculation, creating number hypothesis, and laying out the proverbial strategy that supports current math. Crafted by these early mathematicians laid the foundation for future headways, with their standards staying indispensable to contemporary numerical hypothesis.

During the Islamic Brilliant Age (eighth to fourteenth hundred years), researchers in the Center East and North Africa encouraged Greek numerical information, protecting and developing it. Mathematicians like Al-Khwarizmi, whose name is the foundation of the expression "calculation," and Omar Khayyam made progresses in variable based math, geometry, and mathematical examination. Their work was subsequently sent to Europe, where it helped flash the Renaissance and the resulting Logical Transformation.

The Edification period achieved the improvement of analytics by Isaac Newton and Gottfried Wilhelm Leibniz, changing science into an amazing asset for logical request. Throughout the long term, math kept on advancing, spreading into various subfields, including geography, number hypothesis, and unique polynomial math. This authentic excursion mirrors the discipline's dynamic nature and its significant effect on how we might interpret the world.

Howard's Underlying Experience with Arithmetic

Terrence Howard's excursion into the domain of arithmetic is just about as unpredictable as his methodology. Dissimilar to numerous who follow a customary scholastic way, Howard's advantage in numbers and conditions rose up out of a mix of individual encounters and scholarly interest. Experiencing childhood in a difficult climate, Howard tracked down comfort in the conviction and design of science. Numbers, in contrast to the mayhem of his environmental factors, adhered to guidelines and examples that he could foresee and control.

Howard's interest with science developed during his school years, where he experienced formal numerical speculations that both captivated and confused him. He frequently ended up scrutinizing the acknowledged standards and techniques, feeling that there was something generally wrong. This wariness, instead of dissuading him, powered his longing to profoundly investigate and grasp the fundamental standards of math more.

One significant second in Howard's numerical excursion was his experience with crafted by

Charles Sanders Peirce, a rationalist and mathematician known for his commitments to rationale and semiotics. Peirce's thoughts regarding the idea of numerical thinking resounded with Howard, empowering him to dive further into the philosophical parts of the discipline. This investigation drove him to scrutinize the groundworks of math and to foster his novel point of view regarding the matter.

Ordinarily Acknowledged Numerical Standards

Present day math is based upon a bunch of crucial standards and maxims that structure the reason for additional investigation and improvement. These standards incorporate the idea of numbers, the thought of numerical tasks, and the guidelines overseeing numerical rationale and thinking. At the core of these standards lies the conviction that science is a predictable and intelligent framework, where every assertion can be gotten from a bunch of essential maxims.

One of the most primary standards is the idea of numbers. Numbers are the structure blocks of science, addressing amounts and connections in an exact and extract way. The improvement of various number frameworks, like regular numbers, numbers, judicious numbers, and genuine numbers, has empowered mathematicians to depict and dissect a great many peculiarities.

Numerical tasks, like expansion, deduction, augmentation, and division, are the apparatuses used to control numbers and reveal connections between them. These activities keep explicit guidelines and properties, like commutativity, associativity, and distributivity, which guarantee consistency and consistency in numerical thinking.

Another key guideline is the utilization of maxims and hypotheses. Aphorisms are central insights that are acknowledged without evidence, filling in as the establishment for building numerical hypotheses. Hypotheses, then again, are articulations that are demonstrated in light of these maxims and recently settled hypotheses. This progressive design permits mathematicians to develop mind boggling and unpredictable frameworks of thought, where each new disclosure expands upon crafted by past ages.

Howard's Reactions

Terrence Howard's study of current arithmetic bases on a few key regions that he accepts are defective or fragmented. One of his essential reactions is the dependence on adages that are acknowledged undeniably. Howard contends that these fundamental bits of insight, while advantageous, may not necessarily in all cases hold up under a magnifying glass. He accepts that a more careful assessment of these maxims is important to guarantee the honesty of numerical thinking.

One more area of dispute for Howard is the deliberation innate in current arithmetic. He fights that the move towards progressively conceptual ideas and hypotheses, while mentally invigorating, frequently prompts a distinction from useful reality. Howard advocates for a more grounded way to deal with science, one that remains intently attached to perceptible peculiarities and genuine applications.

Howard additionally disagrees with the unbending nature of numerical schooling. He contends that the ongoing framework, with its accentuation on repetition learning and remembrance, smothers inventiveness and decisive reasoning. As indicated by Howard, understudies are frequently educated to acknowledge numerical standards without figuring out their fundamental reasoning, prompting a shallow handle of the subject. He requires a more unique and intelligent way to deal with showing science, one that energizes investigation and free thought.

Besides, Howard questions the culmination of current numerical speculations. He accepts that there are holes and irregularities in how we might interpret central ideas, like the idea of numbers and the way of behaving of numerical activities. Howard's investigate reaches out to the field of analytics, where he challenges the regular translations of cutoff points and infinitesimals. He proposes elective perspectives that he trusts offer a more precise portrayal of numerical reality.

Howard's Recommendations for Development

In light of these reactions, Terrence Howard offers a few proposition for working on the field of math. One of his key ideas is the reconsideration of numerical adages. Howard trusts that by returning to and thoroughly testing these central bits of insight, mathematicians can guarantee that their hypotheses are based on a strong and dependable premise.

Howard likewise advocates for a more integrative way to deal with science, one that overcomes any barrier between unique hypothesis and down to earth application. He recommends that mathematicians ought to zero in on creating ideas and techniques that have direct significance to certifiable issues, as opposed to chasing after reflection for the good of its own. This methodology, he contends, wouldn't just upgrade the utility of science yet additionally make it more open and drawing in for a more extensive crowd.

In the domain of schooling, Howard requires a shift towards experiential learning. He accepts that understudies ought to be urged to investigate numerical ideas through involved exercises and true models, as opposed to only remembering equations and methodology. This technique, he contends, would cultivate a more profound comprehension of science and rouse more noteworthy inventiveness and development.

At long last, Howard underscores the significance of interdisciplinary coordinated effort. He accepts that science shouldn't exist in disconnection yet ought to be coordinated with different fields of study, like physical science, science, and theory. By cooperating, researchers from various disciplines can acquire new experiences and foster a more all encompassing comprehension of the world.

CHAPTER 2:

Identifying the Flaws

Generally Acknowledged Numerical Standards
Present day math is grounded in a bunch of standards and sayings that are acknowledged as undeniable bits of insight. These incorporate the essential activities of number-crunching, the properties of numbers, and the primary ideas of calculation and variable based math. These standards structure the bedrock whereupon more mind boggling numerical speculations are

fabricated, directing mathematicians in their mission to depict and figure out the universe.

One basic idea is that of numbers and their activities. The regular numbers (1, 2, 3, ...) and their expansions into whole numbers, sane numbers, genuine numbers, and complex numbers give a structure to measuring and breaking down the world. The guidelines of number-crunching tasks — expansion, deduction, increase, and division — are taken as given, with properties like commutativity ($a + b = b + a$), associativity (($a + b) + c = a + (b + c)$), and distributivity ($a(b + c)$ = stomach muscle + ac) being generally acknowledged.

Another essential standard is Euclidean calculation, which portrays the properties of room and shapes in view of a bunch of hypothesizes illustrated by the old Greek mathematician Euclid. These proposes, for example, the possibility that a straight line can be drawn between any two focuses and that okay points are equivalent, structure the reason for a lot of traditional calculation.

In variable based math, the thought of addressing conditions and understanding capabilities and their ways of behaving is significant. The advancement of logarithmic designs like gatherings, rings, and fields has permitted mathematicians to extract and sum up these ideas, prompting significant bits of knowledge and applications in different areas of science and designing.

Howard's Reactions of Numerical Standards

Terrence Howard's scrutinize of present day math is established in his conviction that a portion of these essential standards are imperfect or fragmented. He challenges the tried and true way of thinking that these sayings and rules are reliable, contending that a nearer assessment uncovers irregularities and holes that should be tended to.

1. The Nature of Numbers:
Howard questions the customary comprehension of numbers and their properties. He contends that the ongoing numeric framework, while utilitarian, may not completely catch the intricacies of the real world. For instance, he focuses to the treatment of silly numbers, for example, π (pi) and $\sqrt{2}$ (the square foundation of 2), which can't be communicated as precise divisions. Howard proposes that the dependence on such approximations shows a more profound issue with our conceptualization of numbers.

2. Arithmetic Operations:
Howard additionally scrutinizes the essential activities of number-crunching. He contends that the principles administering expansion, deduction, duplication, and division depend on suppositions that may not necessarily turn out as expected. For example, he challenges the distributive property ($a(b + c)$ = stomach muscle + ac), recommending that there are circumstances where this standard separates. Howard's elective methodology includes

reexamining these tasks from a more central point of view, possibly prompting new bits of knowledge and strategies.

3. Geometric Postulates:
In the domain of calculation, Howard disagrees with a portion of Euclid's hypothesizes. He contends that these hypothesizes, while valuable for depicting romanticized shapes and spaces, don't necessarily in every case line up with the intricacies of this present reality. For instance, the thought that a straight line can be drawn between any two focuses expects a degree of accuracy that may not be reachable practically speaking. Howard advocates for a more adaptable and versatile way to deal with calculation that better mirrors the subtleties of actual space.

4. Algebraic Structures:
Howard's scrutinize reaches out to mathematical designs and the techniques used to settle conditions. He contends that the conventional ways to deal with variable based math, which

depend vigorously on unique images and controls, can cloud the hidden real factors they are intended to address. Howard recommends that a more instinctive and visual way to deal with variable based math could give more clear bits of knowledge and make the subject more open.

Explicit Regions Howard Views as Hazardous

1. Limitations of Calculus:
Howard is especially reproachful of the underpinnings of math, particularly the ideas of cutoff points and infinitesimals. He contends that the customary meanings of these ideas are excessively conceptual and don't sufficiently catch the real essence of persistent change. Howard proposes a reconsidering of math that centers more around substantial portrayals and natural comprehension.

2. Complex Numbers and Nonexistent Units:

One more area of dispute for Howard is the utilization of mind boggling numbers and nonexistent units. He questions the need of presenting fanciful numbers (I, where $i^2 = -1$) to tackle specific conditions, contending that this approach entangles as opposed to explains numerical comprehension. Howard proposes investigating elective techniques for managing issues that customarily require complex numbers.

3. Abstract Algebra:
Howard additionally evaluates the theoretical idea of present day variable based math, especially the investigation of logarithmic designs like gatherings, rings, and fields. He contends that the emphasis on deliberation can prompt a distinction from functional applications and certifiable issues. Howard advocates for a more grounded way to deal with polynomial math that remains intently attached to perceptible peculiarities and regular encounters.

4. Mathematical Education:

Howard is vocal about the deficiencies of numerical training, especially the accentuation on repetition remembrance and procedural information. He accepts that this approach smothers inventiveness and decisive reasoning, prompting a shallow comprehension of numerical ideas. Howard requires a more unique and intelligent way to deal with showing math, one that energizes investigation and free thought.

Models and Delineations

To show his focuses, Howard frequently utilizes basic yet provocative models that challenge the customary way of thinking. For example, he could request understudies to rethink the essential activities from math by investigating elective approaches to consolidating numbers. This could include visual portrayals, like utilizing mathematical shapes or actual items, to give a more natural comprehension of numerical standards.

In math, Howard could show the restrictions of Euclidean hypothesizes by looking at certifiable situations where these principles don't hold. For instance, he could investigate the way of behaving of lines and points in bended or sporadic spaces, featuring the requirement for a more adaptable and versatile way to deal with mathematical thinking.

Howard's investigate of analytics could include reexamining the idea of cutoff points and infinitesimals utilizing visual and substantial portrayals. By zeroing in on natural seeing as opposed to extract definitions, he plans to make analytics more available and significant to understudies.

The More extensive Ramifications of Howard's Evaluate

Terrence Howard's evaluate of current math has more extensive ramifications for the field overall. By testing central standards and offering elective viewpoints, he urges mathematicians to rethink their suppositions and investigate new roads of request. Howard's methodology underlines the significance of addressing laid out standards and staying open to novel thoughts, even in a field as generally unbending as math.

Besides, Howard's emphasis on making arithmetic more available and pertinent to regular day to day existence resounds with more extensive instructive and cultural objectives. By advancing a more natural and useful way to deal with numerical thinking, he plans to motivate more noteworthy interest and commitment to the subject, especially among understudies who might feel distanced by customary techniques.

CHAPTER 3:

The Issue with Numbers

Traditional Appreciation of Numbers

Numbers are the supporting of math, tending to sums, assessments, and associations. The traditional understanding of numbers encompasses a couple of unmistakable sorts, each with its stand-out properties and applications:

1. Natural Numbers: The course of action of positive numbers (1, 2, 3, ...) used for counting and mentioning.
2. Whole Numbers: Ordinary numbers including zero (0, 1, 2, 3, ...).
3. Integers: Whole numbers including their negative accomplices (..., - 3, - 2, - 1, 0, 1, 2, 3, ...).
4. Rational Numbers: Numbers that can be imparted as a little piece of two entire numbers (e.g., 1/2, - 3/4, 5).
5. Irrational Numbers: Numbers that can't be imparted as an insignificant piece of two numbers (e.g., $\sqrt{2}$, π).

6. Real Numbers: All wise and senseless numbers joined.

7. Complex Numbers: Numbers that consolidate a veritable part and a whimsical part (e.g., 3 + 4i, where I is the nonexistent unit with the property $i^2 = -1$).

These number systems structure the bedrock of mathematical exercises, from fundamental math to bleeding edge examination. Numbers are treated as hypothetical components that keep express rules and associations, enabling mathematicians to exhibit and deal with many issues.

Howard's Perspective on Numbers

Terrence Howard's investigation of numbers begins from his conviction that the standard appreciation and depiction of numbers are inherently blemished. He battles that our continuous numeric system, while valuable, doesn't totally get the complexities and nuances of this present reality. Howard's perspective is impacted by his remarkable experiences and his yearning to explore elective viewpoints about math.

1. Questioning the Reality of Irrational Numbers:
Howard is particularly dubious of unreasonable numbers. These numbers, which can't be conveyed as a clear division, are often tended to as non-reiterating, non-finishing decimals. For example, the number π is around 3.14159, but its decimal improvement immeasurably without repeats. Howard battles that the presence of such numbers suggests a level of imprecision and consultation that is in struggle with the significant reality we experience.

2. Reevaluating the Possibility of Nothing and Infinity:
Howard moreover questions the thoughts of nothing and perpetuation. Zero, tending to the

deficit of sum, is a vital part in our number system. Regardless, Howard fights that zero is a determined create instead of a significant reality. Basically, he fights that boundlessness, oftentimes used to depict unbounded sums or endpoints, is a hypothetical believed that doesn't connect with anything concrete in the genuine world.

3. Alternative Numeric Systems:

In his excursion to find a more exact depiction of this present reality, Howard has examined elective numeric systems. He has proposed the use of "terronian" numbers, a system that he confides in better gets the associations and properties of sums as a general rule. This structure incorporates reexamining the fundamental exercises of extension, derivation, duplication, and division, and may offer new encounters into mathematical associations.

Explore of the Continuous Numeric Depictions and Assessments

1. Precision and Approximations:
One of Howard's key assesses is the reliance on approximations in our continuous numeric structure. Senseless numbers, for example, are habitually approximated for realistic purposes, inciting a lack of exactness. Howard battles that this reliance on approximations shows a focal deficiency by they way we could decipher numbers and their properties.

2. Abstract versus Concrete:
Howard's concentrate in like manner focuses on the hypothetical thought of customary numeric depictions. He acknowledges that the pondering characteristic in numbers can provoke a qualification from rational reality. For instance, while the possibility of absurd numbers is mathematically genuine, it may not really in all cases line up with the significant experiences and assessments we experience as a general rule. Howard advocates for a more concrete and

natural method for managing numbers, one that remains eagerly joined to recognizable idiosyncrasies.

3. Mathematical Undertakings and Their Properties:
Howard questions the authenticity of the properties regulating mathematical undertakings. For example, the distributive property (a(b + c) = stomach muscle + ac) is a groundwork of number shuffling, but Howard suggests that there may be conditions where this standard doesn't go out true to form. He battles that a more thorough evaluation of these errands is vital to ensure that they exactly reflect the associations and approaches to acting of sums as a general rule.

Elective Viewpoints and Approaches

1. Visual and Numerical Representations:

One of Howard's proposed decisions incorporates using visual and numerical depictions to sort out numbers and their associations. By focusing in on concrete and regular models, he acknowledges that we can secure an all the more clear and more precise understanding of mathematical thoughts. For example, as opposed to relying upon dynamic pictures, Howard proposes using shapes and graphs to address numbers and exercises, making the associations between them more unquestionable and justifiable.

2. Revisiting Focal Axioms:

Howard advocates for a reexamination of the key proverbs that help our numeric structure. By tending to and completely testing these fundamental assumptions, he acknowledges that we can uncover anticipated flaws and encourage an all the more remarkable and strong framework for math. This association incorporates testing long-held convictions and being accessible to noteworthy considerations and perspectives.

3. Integrative and Interdisciplinary Approaches: Howard highlights the meaning of consolidating pieces of information from various disciplines, as actual science, science, and thinking, into the examination of number juggling. By embracing an interdisciplinary strategy, he acknowledges that we can get an all the more sweeping understanding of numbers and their properties. This coordination can incite new procedures and thoughts that better mirror the complexities of this current reality.

Genuine Repercussions

1. Engineering and Technology:
The exactness and accuracy of numeric depictions are essential in fields like planning and development. Howard's examine suggests that by developing more definite and regular numeric systems, we can chip away at the steady quality and ampleness of mechanical game plans. For example, reexamining the depiction of

nonsensical numbers could provoke more precise calculations in planning applications.

2. Scientific Research:
In legitimate investigation, the precision of assessments and calculations is head. Howard's elective viewpoints on numbers could really provoke new procedures for driving preliminaries and analyzing data. By laying out numeric depictions in perceivable characteristics, analysts could achieve more exact and huge results.

3. Education:
Howard's concentrate in like manner has immense implications for mathematical tutoring. By embracing a more regular and visual method for managing numbers, educators can make the subject more open and attracting for students. This shift could empower a more significant cognizance of mathematical thoughts and energize more critical interest and creative mind in the field.

CHAPTER 4:

Geometry and Spatial Understanding

Customary Standards of Math

Calculation, one of the most seasoned parts of arithmetic, manages the properties and connections of focuses, lines, surfaces, and solids. It is established in crafted by old mathematicians, for example, Euclid, who organized the field in his original work, "Components." Euclidean calculation, in light of a bunch of five proposes, depicts the way of behaving of shapes and spaces in a level, two-layered plane.

The five Euclidean hypothesizes are:
1. A straight line section can be drawn joining any two focuses.

2. Any straight line fragment can be broadened endlessly in an orderly fashion.

3. A circle can be drawn with some random place and span.

4. Okay points are compatible.

5. On the off chance that a straight line falling on two straight lines makes the inside points on a similar side under two right points, then the two straight lines, whenever broadened endlessly, meet on that side on which the points are under two right points.

These proposes structure the establishment for quite a bit of old style calculation, permitting mathematicians to determine properties of shapes, points, and distances. Over the long run, math has extended to incorporate non-Euclidean calculations, for example, exaggerated and elliptic calculation, which investigate spaces that don't stick to Euclid's fifth propose. These progressions have prompted huge improvements in science, physical science, and different fields.

Howard's Reactions of Customary Math

Terrence Howard's scrutinize of customary math centers around its constraints and the requirement for a more adaptable and versatile methodology. He trusts that the unbending design of Euclidean math, while valuable, doesn't completely catch the intricacies of this present reality. Howard's point of view difficulties the basic hypothesizes and proposes elective approaches to figuring out spatial connections.

1. Questioning Euclid's Postulates:
Howard contends that Euclid's hypothesizes, especially the thought of boundlessly expanding lines and amazing circles, are admirations that don't necessarily in every case line up with actual reality. He calls attention to that in reality, lines and circles are dependent upon blemishes and requirements. By scrutinizing these romanticized ideas, Howard looks to foster a more reasonable and pragmatic way to deal with calculation.

2. The Nature of Space:
Howard battles that conventional calculation doesn't enough record for the dynamic and bended nature of room. He proposes that the standards of Euclidean math are restricted when applied to the intricacies of the actual universe, which is much of the time better portrayed by the ebb and flow and bends seen in non-Euclidean calculations. Howard advocates for a more nuanced comprehension of room that consolidates these components.

3. Flexibility and Adaptability:
Howard underlines the requirement for an adaptable and versatile way to deal with math that can oblige different settings and applications. He contends that the unbending design of customary calculation can block imagination and breaking point the investigation of groundbreaking thoughts. By taking on a more adaptable structure, mathematicians can more readily address the different difficulties and peculiarities experienced in reality.

Elective Ways to deal with Math

Howard proposes a few elective ways to deal with math that intend to address the restrictions of customary techniques and proposition new experiences into spatial connections.

1. Visual and Instinctive Methods:
One of Howard's key recommendations is the utilization of visual and instinctive strategies to grasp mathematical ideas. He accepts that by zeroing in on substantial portrayals and visual models, we can acquire a more clear and more exact comprehension of shapes and spaces. This approach includes utilizing outlines, actual models, and intuitive apparatuses to investigate mathematical connections and properties.

CHAPTER 5:

Algebra and Equations

Conventional Standards of Polynomial math

Variable based math is a part of science that arrangements with images and the guidelines for controlling those images to tackle conditions and depict connections. The starting points of variable based math can be followed back to antiquated civic establishments, including the Babylonians and Greeks, however it was crafted by Persian mathematician Al-Khwarizmi that established the groundwork for the deliberate investigation of variable based math. His name is the base of the expression "calculation," featuring the significance of mathematical strategies in critical thinking.

The conventional standards of variable based math include:

1. Variables and Constants: Factors address obscure amounts that can change, while constants are fixed qualities. Together, they structure the reason for mathematical articulations and conditions.

2. Operations and Properties: Mathematical activities incorporate expansion, deduction, augmentation, and division. These activities follow explicit properties like commutativity, associativity, and distributivity.

3. Equations and Inequalities: Conditions are explanations that affirm the fairness of two articulations, while disparities depict the connection between articulations that are not be guaranteed to approach. Addressing conditions and imbalances includes finding the upsides of factors that fulfill these circumstances.

4. Functions and Graphs: Capabilities depict the connection among info and result values, frequently addressed graphically. Understanding the way of behaving of capabilities is a critical part of polynomial math.

5. Polynomials and Factoring: Polynomials are articulations including factors and coefficients.

Considering is the method involved with separating polynomials into easier parts, which is fundamental for settling more significant level conditions.

These standards give the establishment to further developed subjects in polynomial math, like direct polynomial math, dynamic variable based math, and arithmetical calculation.

Howard's Reactions of Customary Polynomial math

Terrence Howard's investigate of conventional variable based math fixates on its theoretical nature and the likely detach between arithmetical techniques and viable applications. He contends that while polynomial math is an amazing asset for taking care of numerical issues, its dependence on conceptual images and rules can some of the time dark the fundamental reality.

1. Abstract Images and Manipulations:

Howard battles that the weighty dependence on dynamic images in polynomial math can prompt a shallow comprehension of numerical ideas. He accepts that understudies frequently figure out how to control images without completely getting a handle on the significance behind them. This can bring about a distinction between mathematical procedures and their certifiable applications.

2. Overemphasis on Procedural Knowledge:
Howard condemns the accentuation on procedural information in variable based math training, where understudies are educated to follow explicit moves toward settle conditions without figuring out the fundamental standards. He contends that this approach smothers inventiveness and decisive reasoning, as understudies become zeroed in on remembering systems as opposed to investigating the rationale and thinking behind them.

3. Complexity and Accessibility:

Howard brings up that the theoretical idea of variable based math can make it distant to numerous understudies. The utilization of mind boggling images and documentation can be scary, prompting disappointment and separation. He advocates for a more instinctive and visual way to deal with variable based math that makes the subject more open and locking in.

Elective Ways to deal with Polynomial math

Howard proposes a few elective ways to deal with polynomial math that intend to address its theoretical nature and upgrade understanding and openness.

1. Concrete and Visual Representations:
One of Howard's key recommendations is to utilize concrete and visual portrayals to show logarithmic ideas. By utilizing actual items, charts, and intelligent instruments, understudies can foster a more natural comprehension of

factors, tasks, and conditions. This approach underscores the importance behind the images and helps overcome any barrier between dynamic ideas and commonsense applications.

2. Contextual and Genuine Applications:
Howard advocates for showing polynomial math with regards to true applications. By demonstrating the way that logarithmic strategies can be utilized to tackle useful issues, understudies can see the importance and utility of the subject. This approach urges understudies to investigate and apply mathematical ideas in different settings, improving their comprehension and enthusiasm for the subject.

3. Exploratory and Request Based Learning:
Howard underlines the significance of exploratory and request based learning in polynomial math schooling. Rather than zeroing in exclusively on procedural information, he recommends empowering understudies to research and find logarithmic standards through investigation and trial and error. This approach

encourages inventiveness, decisive reasoning, and a more profound comprehension of mathematical ideas.

4. Integrative and Interdisciplinary Approaches: Howard trusts that incorporating variable based math with different disciplines, like physical science, science, and financial matters, can give a more comprehensive comprehension of numerical ideas. By investigating the associations among variable based math and different fields, understudies can foster a more extensive viewpoint and see the interdisciplinary idea of math.

Models and Outlines

To outline his focuses, Howard frequently utilizes reasonable models and visual models that challenge traditional arithmetical techniques. For example, he could utilize mathematical shapes to address arithmetical articulations and show the way that these shapes can be controlled to tackle conditions. This approach assists understudies with fostering a more concrete and natural comprehension of logarithmic ideas.

In showing the idea of factors, Howard could utilize actual items, like blocks or dabs, to address various amounts. By controlling these items, understudies can investigate the connections among factors and constants in an unmistakable and visual manner.

Howard's way to deal with tackling conditions could include utilizing graphs and visual models to address the means in question. For instance, he could utilize an equilibrium scale to show the standard of keeping up with correspondence while tackling straight conditions. This visual portrayal assists understudies with getting a handle on the rationale and thinking behind the means.

The More extensive Ramifications of Howard's Scrutinize

Terrence Howard's scrutinize of customary polynomial math has more extensive ramifications for the field of math and training. By testing the theoretical idea of polynomial math and proposing elective methodologies, he urges instructors and mathematicians to reconsider their presumptions and investigate new techniques for educating and learning.

1. Enhancing Availability and Engagement:
Howard's accentuation on concrete and visual portrayals, certifiable applications, and exploratory learning can make variable based math more open and connecting with for a more extensive scope of understudies. This approach can assist with demystifying the subject and move more noteworthy interest and energy for math.

2. Promoting Decisive Reasoning and Creativity:
By empowering understudies to investigate and find logarithmic standards through request based learning, Howard's methodology cultivates decisive reasoning and innovativeness. Understudies are enabled to think autonomously, question suspicions, and foster their comprehension own might interpret arithmetical ideas.

3. Integrating Science with Other Disciplines:
Howard's integrative and interdisciplinary methodology features the associations among polynomial math and different fields of study. This point of view can prompt a more comprehensive comprehension of science and its applications, advancing joint effort and development across disciplines.

CHAPTER 6:

The Role of Mathematical Education

Models and Outlines

To outline his focuses, Howard frequently utilizes pragmatic models and intelligent exercises that challenge traditional educating strategies. For example, he could utilize involved tests and visual models to show mathematical ideas, permitting understudies to investigate connections and examples in an unmistakable and natural manner.

In math, Howard could utilize actual articles and dynamic programming to show the properties of shapes and spaces. By controlling these items and envisioning their changes, understudies can foster a more profound comprehension of mathematical standards.

Howard's way to deal with showing analytics could include true applications and visual models that represent the ideas of cutoff points, subsidiaries, and integrals. By zeroing in on substantial portrayals and useful issues, understudies can see the importance and utility of math in regular day to day existence.

The More extensive Ramifications of Howard's Investigate

Terrence Howard's evaluate of conventional numerical instruction has more extensive ramifications for the field of schooling and the manner in which we approach educating and learning. By testing traditional strategies and proposing elective methodologies, he urges teachers to reconsider their suppositions and investigate better approaches to motivate and connect with understudies.

1. Transforming Instructive Practices:

Howard's accentuation on exploratory learning, substantial portrayals, and genuine applications can change instructive practices and make arithmetic more available and locking in. This approach can rouse more prominent interest and excitement for the subject, cultivating a deep rooted love of learning.

2. Promoting Value and Inclusion:
By supporting for customized and adaptable educational plans, Howard's methodology can advance value and consideration in math training. By obliging the different necessities and interests of understudies, teachers can establish a more comprehensive learning climate that upholds all students.

3. Fostering Decisive Reasoning and Creativity:
Howard's emphasis on request based learning and interdisciplinary methodologies can encourage decisive reasoning and inventiveness. By empowering understudies to investigate and find numerical ideas, teachers can sustain autonomous idea and development.

4. Preparing Understudies for the Future:
Howard's accentuation on certifiable applications and interdisciplinary associations can all the more likely plan understudies for what's in store. By creating pragmatic critical thinking abilities and a wide comprehension of numerical ideas, understudies can be better prepared to explore the intricacies of the cutting edge world.

Terrence Howard's evaluate of conventional numerical instruction moves us to reevaluate our way to deal with educating and learning science. By scrutinizing the accentuation on procedural information, dynamic ideas, and state sanctioned testing, and proposing elective techniques that underline exploratory learning, substantial portrayals, true applications, and customized educational plans, Howard offers a new point of view on math instruction.

CHAPTER 7:

Mathematical Philosophy

Customary Numerical Way of thinking

Numerical way of thinking investigates the nature, degree, and ramifications of arithmetic. It resolves central inquiries regarding numerical truth, the presence of numerical articles, and the connection among math and reality. Customary numerical way of thinking has been molded by different ways of thinking, including:

1. Platonism:
Platonists accept that numerical articles exist freely of human idea, in a theoretical, non-actual domain. As per this view, numerical insights are found as opposed to concocted. Plato's way of thinking sets that these theoretical structures are more genuine than the actual world, and that mathematicians reveal everlasting realities about these structures.

2. Formalism:

Formalists, for example, David Hilbert, contend that math is an arrangement of images and rules for controlling these images. They view math as a making of the human brain, without any trace of any inherent significance outside its proper design. In this view, numerical proclamations are valid in the event that they can be gotten from maxims utilizing acknowledged rules of derivation.

3. Constructivism:

Constructivists, including mathematicians like L.E.J. Brouwer, declare that numerical articles are built by the human brain instead of existing freely. They stress the significance of useful verifications, which unequivocally show the presence of a numerical item by building it.

4. Logicism:

Logicists, for example, Bertrand Russell and Alfred North Whitehead, intend to diminish arithmetic to rationale. They contend that numerical insights can be gotten from coherent maxims and standards, and that arithmetic is basically an expansion of rationale.

5. Empiricism:

Empiricists, similar to John Stuart Plant, fight that numerical information is gotten from tangible experience. They contend that numerical ideas are deliberations from actual perceptions and that numerical bits of insight are affirmed through observational proof.

Howard's Reactions of Customary Numerical Way of thinking

Terrence Howard's investigate of customary numerical way of thinking difficulties the basic suspicions of these ways of thinking. He contends that ordinary perspectives on arithmetic are excessively inflexible and neglect to catch the dynamic and advancing nature of numerical request.

1. Abstract and Static Nature:
Howard fights that conventional numerical way of thinking, especially Platonism and formalism, sees math as a theoretical and static discipline. He trusts this viewpoint ignores the dynamic and inventive parts of numerical idea. As indicated by Howard, science ought to be viewed as a living, developing field that adjusts to groundbreaking thoughts and revelations.

2. Separation from Reality:
Howard scrutinizes the division among science and reality in conventional way of thinking. He contends that review numerical articles as existing in a theoretical domain (Platonism) or as simple images (formalism) separates arithmetic from the actual world. Howard advocates for a more coordinated view that recognizes the reasonable and observational parts of math.

3. Exclusivity and Elitism:
Howard calls attention to that customary numerical way of thinking can make an elitist

perspective on arithmetic, open just to the individuals who can explore its theoretical and formal designs. He accepts this eliteness restricts the scope and effect of math, and that a more comprehensive and open methodology is required.

4. Underestimation of Human Creativity:
Howard accentuates the job of human imagination and instinct in numerical revelation. He contends that customary methods of reasoning, especially formalism and logicism, underrate the significance of imaginative reasoning in arithmetic. Howard accepts that science isn't simply a mechanical cycle yet in addition a work of art that includes creative mind and development.

Howard's Option Numerical Way of thinking

Howard proposes an option numerical way of thinking that accentuates the dynamic, innovative, and pragmatic parts of arithmetic. His viewpoint looks to overcome any issues between unique hypothesis and genuine application, and to make math more comprehensive and open.

1. Dynamic and Advancing Nature of Mathematics:

Howard sees math as a dynamic and developing field, continually adjusting to novel thoughts and revelations. He accepts that numerical request ought to be available to correction and development, instead of being compelled by inflexible conventional designs. This point of view empowers a more adaptable and exploratory way to deal with science.

2. Integration with Reality:

Howard advocates for a nearer reconciliation among science and the actual world. He accepts that numerical ideas ought to be grounded in exact perceptions and reasonable applications.

This approach features the importance and utility of math in taking care of certifiable issues.

3. Inclusivity and Accessibility:

Howard underscores the significance of making science comprehensive and open to a more extensive crowd. He advocates for instructing strategies that demystify unique ideas and make numerical thoughts more interesting. By zeroing in on substantial portrayals and true settings, Howard means to move a more extensive appreciation for science.

4. Creativity and Intuition:

Howard's way of thinking puts areas of strength for an on imagination and instinct in numerical revelation. He accepts that numerical idea includes something beyond sensible thinking; it likewise requires creative mind and inventive reasoning. This viewpoint energizes a more comprehensive perspective on math, perceiving the imaginative and natural parts of the discipline.

Models and Delineations

To show his philosophical viewpoint, Howard frequently utilizes models that challenge regular perspectives on science and feature its dynamic and imaginative nature.

1. Revisiting Numerical Constants:
Howard questions the proper idea of numerical constants, like pi (π), contending that these qualities may be surprisingly liquid. By investigating elective portrayals and approximations of these constants, he exhibits the advancing idea of numerical information.

2. Visual and Natural Methods:
Howard advocates for the utilization of visual and natural strategies to investigate numerical ideas. For example, he could utilize mathematical shapes and actual models to address logarithmic articulations, accentuating the association between dynamic images and unmistakable items.

3. Interdisciplinary Connections:
Howard features the interdisciplinary idea of math by investigating its associations with different fields, like craftsmanship, music, and science. He accepts that coordinating math with different disciplines can give new bits of knowledge and cultivate a more profound comprehension of numerical ideas.

The More extensive Ramifications of Howard's Way of thinking

Terrence Howard's option numerical way of thinking has more extensive ramifications for the field of science and instruction. By testing customary perspectives and proposing a more unique, incorporated, and comprehensive methodology, he urges mathematicians and instructors to reexamine their presumptions and investigate better approaches to motivate and connect with understudies.

1. Reinvigorating Numerical Inquiry:

Howard's accentuation on the dynamic and advancing nature of arithmetic can revive numerical request. By empowering receptiveness to groundbreaking thoughts and development, his way of thinking cultivates a more lively and exploratory numerical local area.

2. Bridging the Hole among Hypothesis and Practice:
Howard's coordination of arithmetic with the actual world can overcome any barrier between conceptual hypothesis and reasonable application. This approach features the importance and utility of arithmetic in daily existence and can rouse more noteworthy interest and commitment to the subject.

3. Promoting Inclusivity and Diversity:
Howard's attention on inclusivity and openness can advance more noteworthy variety in the field of arithmetic. By making numerical ideas more

engaging and justifiable, his way of thinking can draw in a more extensive scope of understudies and specialists, improving the discipline with different points of view and thoughts.

4. Encouraging Innovativeness and Innovation:
Howard's acknowledgment of the innovative and instinctive parts of math can empower a more comprehensive perspective on the discipline. By esteeming creative mind and imaginative reasoning, his way of thinking encourages a culture of innovativeness and investigation in numerical exploration and training.

CHAPTER 8:

Real-World Implications

The Commonsense Effect of Numerical Ideas

Math is in many cases seen as a theoretical field, yet its standards have expansive ramifications in different certifiable settings. From designing and financial matters to science and sociologies, numerical ideas assume a vital part in taking care of useful issues and illuminating dynamic cycles. Understanding this present reality ramifications of science can overcome any barrier between hypothetical information and functional application.

1. Engineering and Technology:
 Arithmetic is fundamental to designing and innovation. Ideas like math, straight polynomial math, and differential conditions are fundamental for planning structures, enhancing

frameworks, and growing new innovations. For instance, math is utilized to break down changing circumstances and anticipate framework conduct, while straight polynomial math is basic to PC illustrations and information handling.

2. Economics and Finance:

Numerical models are critical in financial matters and money for breaking down market patterns, assessing speculation potential open doors, and overseeing risk. Ideas like likelihood hypothesis, factual examination, and advancement are utilized to go with informed monetary choices and estimate financial circumstances. For example, monetary models depend on stochastic cycles to anticipate stock costs and evaluate venture chances.

3. Biology and Medicine:

In science and medication, numerical models help to figure out complex natural frameworks and cycles. Epidemiological models utilize differential conditions to foresee the spread of sicknesses, while factual techniques are utilized in clinical preliminaries to assess the viability of therapies. Arithmetic likewise helps with breaking down hereditary information and figuring out transformative examples.

4. Social Sciences:
Arithmetic is progressively utilized in sociologies to dissect and decipher information connected with human way of behaving, social elements, and strategy influences. Strategies like relapse examination, network hypothesis, and game hypothesis are applied to concentrate on friendly peculiarities and illuminate dynamic in regions like general wellbeing, schooling, and metropolitan preparation.

Howard's Reactions of This present reality Use of Arithmetic

Terrence Howard's study of customary numerical methodologies stretches out to their certifiable applications. He contends that customary numerical techniques frequently miss the mark in tending to perplexing, dynamic, and setting subordinate issues. Howard's viewpoint underscores the requirement for additional versatile and useful ways to deal with arithmetic.

1. Static and Unbending Models:

Howard scrutinizes the dependence on static and inflexible numerical models that may not precisely catch the intricacies of genuine frameworks. Customary models frequently expect fixed conditions and straight connections, which can restrict their appropriateness to dynamic and non-direct situations. Howard advocates for additional adaptable models that can adjust to changing circumstances and integrate vulnerability.

2. Lack of Context oriented Relevance:

Howard brings up that customary numerical techniques some of the time need context oriented importance, as they may not represent the particular attributes and imperatives of genuine circumstances. He underlines the requirement for numerical methodologies that are customized to the remarkable settings and necessities of various applications.

3. Overemphasis on Abstraction:
Howard contends that the attention on dynamic numerical ideas can remove math from useful critical thinking. He accepts that an overemphasis on hypothetical viewpoints might prompt arrangements that are challenging to carry out or less compelling in true situations. Howard advocates for a more commonsense methodology that coordinates hypothetical experiences with viable contemplations.

4. Exclusivity and Accessibility:
Howard features the issue of eliteness and openness in the utilization of science. He contends that conventional strategies can be difficult to reach to those without cutting edge numerical preparation, restricting the more extensive effect of numerical arrangements. Howard calls for approaches that make

numerical instruments and methods more open and usable for a more extensive crowd.

Elective Ways to deal with Applying Science

Howard proposes a few elective ways to deal with applying science in genuine settings that address the limits of customary techniques and improve their commonsense effect.

1. Dynamic and Versatile Models:
Howard advocates for the improvement of dynamic and versatile numerical models that can all the more likely handle changing circumstances and vulnerabilities. These models consolidate adaptability and responsiveness, taking into account more exact and applicable investigation of intricate frameworks. For instance, specialist based models and versatile calculations can mimic the way of behaving of people and frameworks in advancing conditions.

2. Contextual and Modified Solutions:

Howard underlines the significance of fitting numerical answers for the particular setting and necessities of every application. By taking into account the special attributes of true issues, mathematicians can foster more significant and compelling arrangements. This approach includes teaming up with space specialists and integrating relevant information into numerical models.

3. Pragmatic and Integrative Approaches:
Howard advocates for a down to earth approach that coordinates hypothetical bits of knowledge with reasonable contemplations. This includes joining numerical thoroughness with useful execution, guaranteeing that arrangements are both hypothetically sound and possible in genuine situations. Interdisciplinary cooperation and true testing are key parts of this methodology.

4. Inclusive and Open Tools:
To resolve issues of restrictiveness and openness, Howard proposes the advancement of

comprehensive and easy to understand numerical apparatuses and assets. This incorporates making instructive materials, programming, and applications that are open to people with differing levels of numerical mastery. By making numerical apparatuses more congenial, Howard plans to expand their effect and convenience.

Models and Representations

Howard frequently utilizes explicit guides to delineate his elective ways to deal with applying arithmetic in certifiable settings.

1. Dynamic Frameworks Modeling:
To address the restrictions of static models, Howard could utilize dynamic frameworks demonstrating to break down complex peculiarities, for example, environmental change or metropolitan traffic designs. By consolidating ongoing information and versatile calculations, these models can give more precise and significant bits of knowledge.

2. Customized Monetary Models:

In finance, Howard could show the advancement of altered monetary models that record for explicit economic situations and speculation objectives. By fitting models to individual or authoritative necessities, these methodologies can give more significant and successful monetary techniques.

3. Pragmatic Designing Solutions:

Howard could feature designing tasks that coordinate hypothetical arithmetic with useful imperatives. For example, planning structures that record for natural elements and material restrictions includes consolidating numerical examination with functional designing contemplations.

4. Inclusive Instructive Resources:

Howard could feature drives that make available numerical instructive assets, like intelligent learning stages and visual guides.

These assets plan to make numerical ideas more reasonable and drawing in for a different crowd.

The More extensive Ramifications of Howard's Points of view

Terrence Howard's points of view on this present reality utilization of arithmetic have more extensive ramifications for how science is utilized and seen. By supporting for dynamic, logical, and comprehensive methodologies, Howard empowers a more reasonable and effective utilization of numerical information.

1. Enhancing Commonsense Relevance:
Howard's accentuation on powerful and versatile models upgrades the useful pertinence of arithmetic. By creating models that can adjust to evolving conditions, mathematicians can give more exact and significant answers for true issues.

2. Improving Openness and Usability:

By pushing for comprehensive and open numerical devices, Howard advances more extensive commitment with science. This approach can democratize numerical information and engage people and associations to successfully apply numerical arrangements.

3. Fostering Interdisciplinary Collaboration:
Howard's emphasis on relevant and tweaked arrangements features the significance of interdisciplinary cooperation. By working with specialists from different fields, mathematicians can foster arrangements that are both hypothetically sound and for all intents and purposes material.

4. Encouraging Advancement and Practicality:
Howard's logical methodology supports advancement and common sense in numerical critical thinking. By coordinating hypothetical experiences with functional contemplations, mathematicians can foster more viable and possible answers for complex issues.

CHAPTER 9:

Responses from the Mathematical Community

Starting Responses to Howard's Evaluates

Terrence Howard's studies of customary numerical methodologies and ways of thinking have gotten differed reactions from the numerical local area. His unusual viewpoints challenge well established perspectives and works on, provoking both help and wariness from various areas of the field.

1. Supportive Reactions:
 A few mathematicians and teachers have invited Howard's scrutinizes as an impetus for development and change in numerical schooling and application. These allies contend that his accentuation on powerful, context oriented, and comprehensive methodologies lines up with the

advancing requirements of contemporary society.

- Creative Instructive Practices: Instructors who advocate for moderate showing strategies find Howard's scrutinize of customary teaching method convincing. They support his call for exploratory learning, visual and natural strategies, and context oriented importance, accepting these methodologies can make math seriously captivating and powerful for understudies.

- Functional Applications: Experts in applied math and interdisciplinary fields value Howard's attention on reasonable and versatile models. They see esteem in creating numerical apparatuses and models that are receptive to genuine intricacies and vulnerabilities.

2. Skeptical Reactions:
Then again, a few individuals from the numerical local area have some misgivings of Howard's viewpoints, especially in regards to the

reasonableness and practicality of his proposed changes.

- Worries about Rigor: Pundits contend that Howard's accentuation on versatility and inclusivity could think twice about thoroughness and accuracy that are basic to numerical practice. They stress that getting away from formalism could debilitate the insightful power and unwavering quality of numerical techniques.

- Execution Challenges: Doubters likewise question the plausibility of carrying out Howard's thoughts on a wide scale. They bring up the difficulties in coordinating unique models and context oriented arrangements into existing educational plans and expert works on, refering to worries about the common sense of such changes.

Key Figures and Their Viewpoints

A few conspicuous mathematicians and teachers have said something regarding Howard's evaluates, offering their bits of knowledge and points of view.

1. Advocates for Reform:
- Paul Lockhart: A vocal backer for change in math schooling, Lockhart upholds Howard's require a more unique and drawing in way to deal with educating math. He concurs with Howard's scrutinize of the "drill-and-kill" strategies and underlines the significance of encouraging imagination and instinct in numerical learning.

- Jo Boaler: Known for her work on further developing arithmetic instruction, Boaler reverberates with Howard's emphasis on making science more comprehensive and significant. She advocates for instructing strategies that advance comprehension and critical thinking instead of repetition retention, lining up with Howard's vision.

2. Critics of Howard's Perspectives:

- Andrew Wiles: A famous mathematician known for demonstrating Fermat's Last Hypothesis, Wiles has communicated worries about creating some distance from customary numerical meticulousness. He contends that while flexibility and logical significance are significant, the basic standards of math ought to stay powerful and exact.

- John Nash: In spite of the fact that Nash has died, his work on game hypothesis and arithmetic keeps on affecting the field. His disciples contend that Howard's study could neglect the mind boggling balance among hypothesis and application that Nash and others have laid out in numerical exploration.

Discussions and Conversations

Howard's scrutinizes have started discussions and conversations inside the numerical local area, featuring the different perspectives on how arithmetic ought to develop.

1. Educational Reform:

The discussion on instructive change bases on Howard's idea for more exploratory and comprehensive educating techniques. Defenders contend that customary strategies neglect to connect with understudies and that Howard's methodology could prompt a more significant and important growth opportunity. Rivals, notwithstanding, alert against leaving laid out rehearses that have demonstrated compelling in building areas of strength for an establishment.

2. Practical Applications:

Conversations on commonsense applications center around the attainability of carrying out powerful and versatile models. Allies of Howard's point of view advocate for incorporating genuine intricacy into numerical displaying, while pundits raise worries about the difficulties and asset prerequisites of creating and sending such models.

3. Philosophical Perspectives:

The philosophical discussions encompassing Howard's perspectives address the idea of numerical truth and the job of deliberation. Howard's require a more even minded and logically important methodology challenges customary philosophical perspectives, inciting conversations about the harmony between dynamic hypothesis and useful application.

Instances of Execution and Effect

In light of Howard's studies, a few establishments and people have started trying different things with elective methodologies in science schooling and application.

1. Innovative Instructive Programs:
- Project-Based Learning: A few schools and instructive projects have embraced project-based learning approaches that line up with Howard's accentuation on investigation and context oriented pertinence. These projects urge

understudies to figure out on genuine issues and apply numerical ideas in viable situations.

- Visual and Intelligent Tools: Instructive innovation designers have made clear line of sight and intuitive devices that make unique numerical ideas more open. These instruments line up with Howard's backing for substantial portrayals and natural comprehension.

2. Adaptive Displaying in Practice:
- Environment Modeling: Specialists have created versatile models for environment expectation that integrate constant information and record for dynamic ecological variables. These models mirror Howard's way to deal with incorporating adaptability and context oriented importance into numerical applications.

- Monetary Gamble Management: Monetary organizations are exploring different avenues regarding versatile calculations and altered models to all the more likely oversee risk and answer market variances. These practices reverberation Howard's accentuation on reasonable and setting explicit arrangements.

The Eventual fate of Arithmetic Considering Howard's Studies

Howard's evaluates have opened up new roads for investigating the eventual fate of arithmetic and its job in the public arena. As the numerical local area keeps on drawing in with his thoughts, a few potential improvements might arise.

1. Evolving Instructive Practices:
 The eventual fate of science training might see a shift towards more powerful, exploratory, and comprehensive methodologies. Instructive changes motivated by Howard's investigates could prompt a seriously captivating and useful opportunity for growth, planning understudies to address genuine difficulties.

2. Advancements in Numerical Modeling:

The improvement of versatile and logically significant models might progress numerical applications across different fields. This development could upgrade the capacity of mathematicians and professionals to actually resolve perplexing and dynamic issues.

3. Philosophical Reconsiderations:
The philosophical discussions ignited by Howard's evaluates may prompt a reconsideration of the nature and job of math. This could bring about a more nuanced comprehension of the harmony between hypothesis, deliberation, and pragmatic application.

CONCLUSION

Terrence Howard's investigation of the issues inside science offers a provocative and groundbreaking point of view on a field customarily considered inflexible and dynamic. His studies of regular numerical methodologies — going from the static idea of conventional models to the distinction among hypothesis and true applications — challenge the numerical local area to reexamine primary practices and ways of thinking.

Howard's accentuation on unique, logical, and comprehensive methodologies looks to overcome any issues between hypothetical science and reasonable critical thinking. His investigates feature the impediments of static models and dynamic speculations, supporting for a more versatile and functional utilization of numerical ideas. By zeroing in on the significance and openness of science, Howard energizes a more extensive commitment with the field, planning to make numerical devices and techniques more pertinent to different true settings.

The reactions from the numerical local area uncover a range of responses, from excited help to wary incredulity. Advocates for change consider Howard's plans to be an essential development in math schooling and application, lining up with contemporary requirements and cultivating advancement. Pundits, nonetheless, express worries about keeping up with numerical meticulousness and the possibility of carrying out such changes for an expansive scope.

As the numerical local area keeps on wrestling with Howard's evaluates, the eventual fate of math is probably going to be portrayed by a mix of customary standards and new methodologies. The developing scene might include more powerful and logically important models, imaginative instructive practices, and a reconsidered philosophical comprehension of science.

Eventually, Howard's points of view welcome mathematicians, teachers, and specialists to participate in continuous discourse and reflection about the job and effect of arithmetic. His commitments highlight the significance of ceaselessly adjusting and developing the field to address intricate, genuine difficulties while keeping a harmony between hypothetical profundity and reasonable relevance.

By looking at Howard's experiences and the local area's reactions, we gain a more profound comprehension of the expected pathways for math to stay an essential and extraordinary discipline. The eventual fate of science will probably be formed by a powerful transaction among development and custom, driven by a promise to making numerical information more important, open, and effective in a quickly impacting world.

ACKNOWLEDGEMENTS

Composing this book has been an excursion of investigation and revelation, made conceivable by the help, direction, and bits of knowledge of numerous people. I'm profoundly thankful to the people who contributed their time, skill, and consolation in the interim.

I, first and foremost, need to say thanks to Terrence Howard for his provocative studies and creative viewpoints on arithmetic. His exceptional way to deal with understanding and applying numerical ideas has given a new focal point through which to look at the field and has roused a significant part of the substance in this book.

I stretch out my ardent thanks to the mathematicians, teachers, and researchers who offered their points of view on Howard's scrutinizes and taken part in significant conversations about the eventual fate of math. Your criticism and experiences have enhanced this book and added to a more nuanced investigation of the main things in need of attention.

Exceptional appreciation goes to the teachers and professionals who are at the very front of carrying out creative numerical methodologies and instructive changes. Your devotion to making science more powerful, pertinent, and open is a demonstration of the advancing idea of the field and fills in as a wellspring of motivation for future turns of events.

I'm likewise appreciative to my partners and companions for their help and support all through the creative cycle. Your excitement for the topic and your productive criticism have been priceless in molding this book.

At last, I might want to recognize my family for their unflinching help and persistence. Your comprehension and support have been a steady wellspring of inspiration.

Much thanks to you to each and every individual who has added to this undertaking. Your commitments have made this investigation of arithmetic and its future conceivable, and I'm profoundly keen to your help.

www.ingramcontent.com/pod-product-compliance
Lightning Source LLC
Chambersburg PA
CBHW071945210526
45479CB00002B/820